Fractions, Decimals and Percentages

Dear Student,

Thank you for buying this book. It will help you with your GCSE Maths by showing you how to gain extra marks. Just one of those marks could move you up a grade. Remember that there is only one mark difference between a Grade D and a Grade C.

Most of the double page spreads in the book consist of three sections: 'What You Need to Know', 'Revision Facts' and 'Questions'.

Read 'What You Need to Know' carefully, making sure that it makes sense to you. You may find it helpful to work with someone else so that you can check that each other understands.

Use the 'Revision Facts' as a reminder — you may like to copy these into an exercise book to create a revision guide to use nearer your exam.

Try all the 'Questions', even those that are easy but especially those that appear to be difficult. There is an answer section in the centre that you can pull out so that you can check your work. If you find that you have made a mistake, don't worry but just try to see where you went wrong. You can learn a lot from mistakes.

To answer some of the questions you may need a pair of compasses, a protractor or a calculator.

Good luck,

Andrew Brodie

YOU NEED TO KNOW

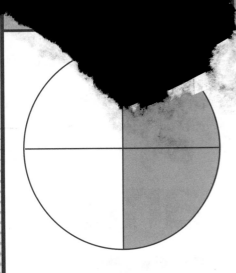

$\frac{2}{4}$ is the same amount as $\frac{1}{2}$

$\frac{2}{4}$ and $\frac{1}{2}$ are equivalent fractions.

$\frac{6}{12}$ is the same amount as $\frac{1}{2}$

$\frac{6}{12}$ and $\frac{1}{2}$ are equivalent fractions.

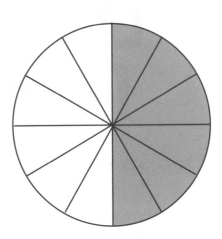

You may be asked to simplify a fraction.
This means that you have to find an equivalent fraction
that has smaller numbers top and bottom.

Example. Simplify $\frac{15}{20}$

To simplify the fraction, divide the top and bottom by the same number:

$$\frac{15}{20} = \frac{3}{4}$$

$\div 5$

$\div 5$

QUESTIONS

Simplify the following fractions:

1 $\frac{35}{42} =$ **2** $\frac{18}{24} =$ **3** $\frac{25}{30} =$

4 $\frac{20}{24} =$ **5** $\frac{16}{18} =$ **6** $\frac{15}{18} =$

7 $\frac{36}{54} =$ **8** $\frac{40}{60} =$ **9** $\frac{75}{100} =$

2

Sometimes we need to do the opposite of simplifying. We may need to alter the denominator to make it match the denominator of another fraction.

The top number is the numerator. → $\frac{1}{2}$ ← The bottom number is the denominator.

Look at these examples:

1. $\frac{1}{2} = \frac{8}{16}$ because $1 \times 8 = 8$
$2 \times 8 = 16$

2. $\frac{3}{4} = \frac{15}{20}$ (× 5 / × 5)

3. $\frac{5}{8} = \frac{35}{56}$ (× 7 / × 7)

4. $\frac{2}{5} = \frac{}{30}$ ←

$\frac{2}{5} = \frac{12}{30}$ ←

You may be asked to find the missing numerator. You can see that the denominator has changed from 5 to 30 so it must have been multiplied by 6. To keep the fractions equivalent you have to multiply the numerator by 6.

REVISION FACTS

✓ Equivalent fractions have the same value.

✓ The top number is the numerator.

✓ The bottom number is the denominator.

✓ Keep fractions equivalent by multiplying or dividing the numerator and denominator by the same number. Remember: do the same to both.

QUESTIONS

Complete the equivalent fractions:

1 $\frac{3}{4} = \frac{}{12}$

2 $\frac{5}{8} = \frac{}{40}$

3 $\frac{10}{50} = \frac{}{5}$

4 $\frac{2}{7} = \frac{}{21}$

5 $\frac{8}{28} = \frac{2}{}$

6 $\frac{3}{9} = \frac{1}{}$

7 $\frac{3}{5} = \frac{}{20}$

8 $\frac{5}{9} = \frac{}{36}$

9 $\frac{1}{2} = \frac{512}{}$

WHAT YOU NEED TO KNOW

$2\frac{1}{2}$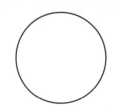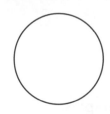

We say that $2\frac{1}{2}$ is a mixed number
because it is a mixture of a whole number part and a fraction part.

You can see that $2\frac{1}{2}$

is the same amount as five halves.

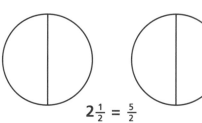 

$2\frac{1}{2} = \frac{5}{2}$

Here are some more:

 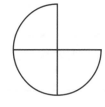  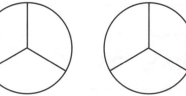

$1\frac{3}{4} = \frac{7}{4}$

 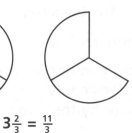

$3\frac{2}{3} = \frac{11}{3}$

Fractions such as $\frac{5}{2}$, $\frac{7}{4}$ and $\frac{11}{3}$ are called improper fractions or top-heavy fractions because they are bigger at the top than at the bottom.

A quick way to make top-heavy fractions:

$2\frac{1}{2}$ $\quad$ $\frac{5}{2}$
$2 \times 2 + 1 = 5$

Here are some more:

$4\frac{5}{8} = \frac{37}{8}$ $\quad$ $4 \times 8 + 5$ $\qquad\qquad$ $6\frac{2}{5} = \frac{32}{5}$ $\quad$ $6 \times 5 + 2$

Sometimes you will have a top-heavy fraction that you need to change to a mixed number.

$$\frac{17}{5}$$ Divide the top by the bottom.

$$17 \div 5 = 3 \text{ r } 2$$

$$\frac{17}{5} = 3\frac{2}{5}$$

Here are some more:

$\frac{13}{3} = 4\frac{1}{3}$ because $13 \div 3 = 4 \text{ r } 1$ $\qquad\qquad$ $\frac{28}{5} = 5\frac{3}{5}$ because $28 \div 5 = 5 \text{ r } 3$

Sometimes the top-heavy fraction can be changed to a whole number:

$\frac{14}{7} = 2$ because $14 \div 7 = 2$ with no remainder.

REVISION FACTS

✓ Numbers like $3\frac{1}{4}$ are called mixed numbers.
✓ Fractions like $\frac{7}{3}$ are called top-heavy fractions.
✓ A mixed number can be changed to a top-heavy fraction by multiplying the denominator by the whole number, then adding the numerator: $2\frac{3}{4} = \frac{11}{4}$
✓ A top-heavy fraction can be changed to a mixed number by dividing the numerator by the denominator: $\frac{17}{3} = 5\frac{2}{3}$

QUESTIONS

Change these mixed numbers to top-heavy fractions:

1 $1\frac{1}{2} =$ $\qquad$ **2** $2\frac{5}{8} =$ $\qquad$ **3** $4\frac{3}{4} =$ $\qquad$ **4** $5\frac{2}{3} =$

5 $3\frac{4}{7} =$ $\qquad$ **6** $2\frac{2}{9} =$ $\qquad$ **7** $5\frac{1}{2} =$ $\qquad$ **8** $12\frac{3}{4} =$

Change these top-heavy fractions to mixed numbers:

9 $\frac{16}{3} =$ $\qquad$ **10** $\frac{15}{4} =$ $\qquad$ **11** $\frac{18}{4} =$ $\qquad$ **12** $\frac{23}{6} =$

13 $\frac{38}{5} =$ $\qquad$ **14** $\frac{27}{4} =$ $\qquad$ **15** $\frac{30}{7} =$ $\qquad$ **16** $\frac{40}{9} =$

WHAT YOU NEED TO KNOW

Look: $\frac{2}{5} + \frac{1}{5} = \frac{3}{5}$ **If the denominators are the same, just add the numerators.**

The denominator does not change.

If the denominators are different you have to make them the same:

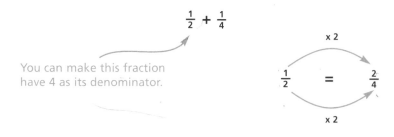

$$\frac{1}{2} + \frac{1}{4}$$

You can make this fraction have 4 as its denominator.

$$\frac{1}{2} = \frac{2}{4}$$

... **so** $\frac{1}{2} + \frac{1}{4} = \frac{2}{4} + \frac{1}{4} = \frac{3}{4}$

Sometimes you need to change both denominators:

$$\frac{1}{2} + \frac{2}{3}$$

Step 1: Find a number that is a multiple of 2 and a multiple of 3: 2 and 3 both fit into 6, so 6 is a multiple of both.

Step 2: Change the first fraction so that its denominator is 6:

$$\frac{1}{2} = \frac{3}{6}$$

Step 3: Change the second fraction so that its denominator is 6:

$$\frac{2}{3} = \frac{4}{6}$$

Step 4: Add the new fractions: $\frac{3}{6} + \frac{4}{6} = \frac{7}{6}$

Step 5: Change to a mixed number if the fraction is top-heavy: $\frac{7}{6} = 1\frac{1}{6}$

WHAT YOU NEED TO KNOW

To subtract fractions you again need to make the denominators match:

$$\frac{3}{4} - \frac{1}{6} = \frac{9}{12} - \frac{2}{12} = \frac{7}{12}$$

4 and 6 both fit into 12.

To make 4 into 12 multiply by 3,
then remember to multiply the numerator by 3.

To make 6 into 12 multiply by 2,
then remember to multiply the numerator by 2.

If you are adding or subtracting mixed numbers, the easiest way is to make both into top-heavy fractions. You may need to simplify the answer.

Examples: 1. $2\frac{3}{4} + 1\frac{1}{2} = \frac{11}{4} + \frac{3}{2} = \frac{11}{4} + \frac{6}{4} = \frac{17}{4} = 4\frac{1}{4}$

2. $3\frac{1}{5} - 2\frac{1}{4} = \frac{16}{5} - \frac{9}{4} = \frac{64}{20} - \frac{45}{20} = \frac{19}{20}$

REVISION FACTS

✓ To add or subtract fractions, made the denominators match.

✓ To add or subtract mixed numbers, change them to top-heavy fractions.

QUESTIONS

1. $\frac{1}{4} + \frac{1}{3} =$

2. $\frac{1}{2} + \frac{3}{5} =$

3. $\frac{2}{3} - \frac{1}{4} =$

4. $\frac{3}{5} - \frac{1}{2} =$

5. $\frac{3}{7} + \frac{2}{5} =$

6. $\frac{2}{3} - \frac{1}{6} =$

7. $1\frac{1}{2} + 3\frac{1}{3} =$

8. $4\frac{1}{4} - 2\frac{1}{7} =$

WHAT YOU NEED TO KNOW

To multiply two fractions, just multiply the numerators and multiply the denominators:

Examples:

1. $\frac{1}{2} \times \frac{3}{4} = \frac{3}{8}$

2. $\frac{2}{5} \times \frac{3}{7} = \frac{6}{35}$

3. $\frac{4}{9} \times \frac{3}{8} = \frac{12}{72}$

This answer can be simplified:

$$\frac{12}{72} = \frac{1}{6}$$

To multiply mixed numbers, change them to top-heavy fractions first.

Examples:

1. $2\frac{3}{4} \times 1\frac{1}{2} = \frac{11}{4} \times \frac{3}{2} = \frac{33}{8} = 4\frac{1}{8}$

2. $1\frac{1}{3} \times 2\frac{4}{5} = \frac{4}{3} \times \frac{14}{5} = \frac{56}{15} = 3\frac{11}{15}$

QUESTIONS

1 $\frac{1}{2} \times \frac{1}{4} =$

2 $\frac{5}{6} \times \frac{3}{4} =$ ⟵ This one can be simplified. Look out for others that can.

3 $\frac{2}{7} \times \frac{3}{8} =$

4 $\frac{5}{9} \times \frac{3}{10} =$

5 $\frac{3}{8} \times \frac{2}{3} =$

6 $\frac{7}{11} \times \frac{2}{3} =$

7 $1\frac{1}{2} \times 1\frac{1}{4} =$

8 $4\frac{3}{8} \times 2\frac{2}{3} =$

9 $5\frac{1}{4} \times 2\frac{4}{7} =$

WHAT YOU NEED TO KNOW

To divide one fraction by another, turn the second one upside down and multiply instead.

Look: $\quad \frac{1}{2} \div \frac{1}{4} = \frac{1}{2} \times \frac{4}{1} = \frac{4}{2} = 2$

Turn the second fraction upside down.

To divide mixed numbers, change them to top-heavy fractions first.

Example: $\quad 2\frac{3}{4} \div 1\frac{1}{3} = \frac{11}{4} \div \frac{4}{3} = \frac{11}{4} \times \frac{3}{4} = \frac{33}{16} = 2\frac{1}{16}$

Change to top-heavy.

Turn 2nd fraction upside down, then multiply.

Change to a mixed number.

REVISION FACTS

✓ To multiply fractions, just multiply numerators and multiply denominators.

✓ To divide fractions, turn the second fraction upside down then multiply.

✓ To multiply or divide mixed numbers, change them to top-heavy fractions first.

QUESTIONS

1 $\quad \frac{3}{7} \div \frac{1}{3} =$

2 $\quad \frac{5}{8} \div \frac{3}{4} =$

3 $\quad \frac{7}{9} \div \frac{2}{3} =$

4 $\quad \frac{3}{4} \div \frac{1}{2} =$

5 $\quad 3\frac{2}{3} \div 2\frac{1}{4} =$

6 $\quad 5\frac{1}{2} \div 1\frac{2}{5} =$

WHAT YOU NEED TO KNOW

To change a fraction to a decimal divide the numerator by the denominator.

Examples:

1. $\frac{1}{2}$ = 1 ÷ 2 = 0.5

$$2 \overline{)1.0}^{\,0.5}$$

2. $\frac{3}{4}$ = 3 ÷ 4 = 0.75

$$4 \overline{)3.00}^{\,0.75}$$

You can use your calculator.

Remember, to change any fraction to a decimal:

it's the top divided by the bottom.

⬆ numerator   ⬆ denominator

Look at these examples:

1. $\frac{1}{3}$ = 1 ÷ 3

If you press   on the calculator

you will get this answer on the screen: 0.333333333

This is called a recurring decimal. If the calculator screen was big enough the 3's would go on forever.

To show this you write: 0.3̇   ⟵ Write the dot clearly over the 3.

2. $\frac{2}{11}$ = 2 ÷ 11

The calculator shows: 0.181818181

The recurring decimal is shown like this: 0.1̇8̇

so $\frac{2}{11}$ = 0.1̇8̇

3. $\frac{4}{37} = 4 \div 37$

The calculator shows: 0.108108108

The recurring decimal is shown like this: $0.\dot{1}0\dot{8}$

$$\frac{4}{37} = 0.\dot{1}0\dot{8}$$

4. $\frac{1}{7} = 1 \div 7$

The calculator shows: 0.142857142

This is where the pattern starts repeating.

$$\frac{1}{7} = 0.\dot{1}4285\dot{7}$$

The dots are shown over the first and last digits of the repeating pattern.

REVISION FACTS

✓ To change a fraction to a decimal: top ÷ bottom.

✓ To show recurring decimals draw a dot over the first and last digits of the repeating pattern.

QUESTIONS

Change these fractions to decimals:

1 $\frac{4}{5} =$

2 $\frac{1}{4} =$

3 $\frac{3}{8} =$

4 $\frac{2}{3} =$

5 $\frac{7}{12} =$

6 $\frac{1}{6} =$

7 $\frac{5}{6} =$

8 $\frac{3}{4} =$

9 $\frac{23}{100} =$

10 $\frac{5}{8} =$

11 $\frac{3}{7} =$

12 $\frac{5}{9} =$

WHAT YOU NEED TO KNOW

To change a fraction to a percentage,
first change it to a decimal then multiply by 100.

Examples:

1. $\frac{1}{2}$ = 0.5 = 50%

2. $\frac{1}{4}$ = 0.25 = 25% 3. $\frac{3}{4}$ = 0.75 = 75%

4. $\frac{1}{7}$ = 0.$\dot{1}$4285$\dot{7}$ = 14.2857% approximately

But, you would normally express percentages to just one or two decimal places,

so $\frac{1}{7}$ = 0.$\dot{1}$4285$\dot{7}$ = 14.3% approximately.

REVISION FACT

✓ To change a fraction to a percentage: top ÷ bottom, then x 100.

QUESTIONS

Change the following fractions to percentages by changing them to decimals first:

1 $\frac{1}{8}$ =   =   **2** $\frac{4}{5}$ =   =

3 $\frac{1}{4}$ =   =   **4** $\frac{5}{8}$ =   =

5 $\frac{1}{2}$ =   =   **6** $\frac{1}{3}$ =   =

7 $\frac{2}{3}$ =   =   **8** $\frac{3}{4}$ =   =

9 $\frac{7}{20}$ =   =   **10** $\frac{7}{8}$ =   =

WHAT YOU NEED TO KNOW

If you do a test, your result may be given to you as a percentage.

Example: Myra gained 16 marks out of 30 in a biology test. What percentage did she achieve?

Step 1:   Express her score as a fraction: $\frac{16}{30}$

Step 2:   Change the fraction to a decimal: $16 \div 30 = 0.5\dot{3}$

Step 3:   Multiply the decimal by 100. $0.5\dot{3} \times 100 = 53.\dot{3}$

Step 4:   With a percentage score you should round to the nearest whole number: 53%

REVISION FACT

✓ To find a percentage score:
   a) Make a fraction.
   b) Change to a decimal.
   c) Multiply by 100.
   d) Round to the nearest whole number.

QUESTIONS

1   Tariq gained 19 out of 20 in a maths test.
    What was his percentage score?

    ----

2   Eric scored 5 out of 36 in a chemistry test.
    What percentage did he gain?

WHAT YOU NEED TO KNOW

Example 1   A train ticket goes up in price from £54 to £57.
What percentage is the increase?

Step 1:   Find the change in price: £57 - £54 = £3.

Step 2:   Create a fraction: $\dfrac{\text{change}}{\text{original price}} = \frac{3}{54}$

Step 3:   Change the fraction to a decimal:
On the calculator, 3 ÷ 54 gives 0.055555555.

Step 4:   Multiply by 100.
The calculator will now show: 5.555555556

Step 5:   Round the answer to one decimal place:
5.6

The percentage increase is about 5.6%

Example 2   A magazine goes up in price from £2.50 to £3.00.
What is the percentage increase?

Step 1:   The change in price is 50p.

Step 2: $\dfrac{\text{change}}{\text{original price}} = \frac{50}{250}$ These must both be expressed in the same way. Here we have expressed them in pence, but we could have shown them in pounds: $\frac{0.5}{2.5}$

Step 3:   50 ÷ 250 = 0.2

Step 4:   0.2 x 100 = 20

This time we don't need to round the answer to 1 decimal place.
The percentage increase is 20%.

p.2 1. $\frac{5}{6}$ because we can divide the top and bottom by 7.

2. $\frac{3}{4}$ 3. $\frac{5}{6}$ 4. $\frac{5}{6}$ 5. $\frac{8}{9}$ 6. $\frac{5}{6}$

7. Some questions are easier to do in stages:

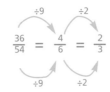

8. $\frac{2}{3}$ 9. $\frac{3}{4}$

p.3 1. $\frac{9}{12}$ because we have multiplied the denominator 4 by 3 to make 12, we must multiply the numerator by 3; this makes 9.

2. $\frac{25}{40}$ 3. $\frac{1}{5}$ 4. $\frac{6}{21}$ 5. $\frac{2}{7}$ 6. $\frac{1}{3}$ 7. $\frac{12}{20}$ 8. $\frac{20}{36}$ 9. $\frac{512}{1024}$

p.5 1. $\frac{3}{2}$ 2. $2\frac{5}{8} = \frac{21}{8}$ Look how we did this one:

3. $\frac{19}{4}$ 4. $\frac{17}{3}$ 5. $\frac{25}{7}$ 6. $\frac{20}{9}$ 7. $\frac{11}{2}$ 8. $\frac{51}{4}$

9. $\frac{16}{3} = 5\frac{1}{3}$ because $16 \div 3 = 5$ r 1 10. $3\frac{3}{4}$ 11. $4\frac{2}{4} = 4\frac{1}{2}$ 12. $3\frac{5}{6}$ 13. $7\frac{3}{5}$

$5 \quad \frac{1}{3}$

14. $6\frac{3}{4}$ 15. $4\frac{2}{7}$ 16. $4\frac{4}{9}$

p.7 1. $\frac{1}{4} + \frac{1}{3} = \frac{3}{12} + \frac{4}{12} = \frac{7}{12}$ 2. $\frac{1}{2} + \frac{3}{5} = \frac{5}{10} + \frac{6}{10} = \frac{11}{10} = 1\frac{1}{10}$

3. $\frac{2}{3} - \frac{1}{4} = \frac{8}{12} - \frac{3}{12} = \frac{5}{12}$ 4. $\frac{3}{5} - \frac{1}{2} = \frac{6}{10} - \frac{5}{10} = \frac{1}{10}$

5. $\frac{3}{7} + \frac{2}{5} = \frac{15}{35} + \frac{14}{35} = \frac{29}{35}$ 6. $\frac{2}{3} - \frac{1}{6} = \frac{4}{6} - \frac{1}{6} = \frac{3}{6} = \frac{1}{2}$

Look: we didn't need to change this denominator as 3 and 6 both fit into 6.

7. $1\frac{1}{2} + 3\frac{1}{3} = \frac{3}{2} + \frac{10}{3} = \frac{9}{6} + \frac{20}{6} = \frac{29}{6} = 4\frac{5}{6}$

8. $4\frac{1}{4} - 2\frac{1}{7} = \frac{17}{4} - \frac{15}{7} = \frac{119}{28} - \frac{60}{28} = \frac{59}{28} = 2\frac{3}{28}$

p.8 1. $\frac{1}{8}$ 2. $\frac{15}{24} = \frac{5}{8}$ 3. $\frac{6}{56} = \frac{3}{28}$ 4. $\frac{15}{90} = \frac{1}{6}$ 5. $\frac{6}{24} = \frac{1}{4}$ 6. $\frac{14}{33}$ 7. $\frac{3}{2} \times \frac{5}{4} = \frac{15}{8} = 1\frac{7}{8}$

8. $\frac{35}{8} \times \frac{8}{3} = \frac{280}{24} = 11\frac{2}{3}$ 9. $\frac{21}{4} \times \frac{18}{7} = \frac{378}{28} = 13\frac{1}{2}$

You may like to simplify this
before changing it to a mixed
number.

p.9 1. $\frac{3}{7} \div \frac{1}{3} = \frac{3}{7} \times \frac{3}{1} = \frac{9}{7} = 1\frac{2}{7}$ 2. $\frac{20}{24} = \frac{5}{6}$ 3. $\frac{21}{18} = 1\frac{1}{6}$ 4. $\frac{6}{4} = 1\frac{1}{2}$

5. $3\frac{2}{3} \div 2\frac{1}{4} = \frac{11}{3} \div \frac{9}{4} = \frac{11}{3} \times \frac{4}{9} = \frac{44}{27} = 1\frac{17}{27}$ 6. $5\frac{1}{2} \div 1\frac{2}{5} = \frac{11}{2} \div \frac{7}{5} = \frac{11}{2} \times \frac{5}{7} = \frac{55}{14} = 3\frac{13}{14}$

In the exam, even if you only get this far,
you should get some marks.

p.11 1. 0.8   2. 0.25   3. 0.375   4. $0.\dot{6}$ 5. $0.58\dot{3}$ 6. $0.1\dot{6}$ 7. $0.8\dot{3}$

8. 0.75   9. 0.23   10. 0.625   11. $0.\dot{4}2857\dot{1}$ Notice the dots because this
decimal repeats like this:
0.428514285714285714...

12. $0.\dot{5}$

p.12 1. 0.125 = 12.5%   2. 0.8 = 80%   3. 0.25 = 25%   4. 0.625 = 62.5%

5. 0.5 = 50%   6. $0.\dot{3}$ = 33.3% because $0.\dot{3}$ = 0.33333 ....

7. $0.\dot{6}$ = 66.6% (or approximately 67%)   8. 0.75 = 75%

9. 0.35 = 35%   10. 0.875 = 87.5%

p.13 1. $\frac{19}{20}$ = 0.95 = 95%   2. $\frac{5}{36}$ = $0.13\dot{8}$ = 13.8%   But we would normally round
this to a whole number, so his
score is approximately 14%.

p.19   1.   Fraction: $\frac{4}{32}$ ←— Change
                         ←— Original value $\frac{4}{32} = \frac{1}{8} = 0.125 = 12.5\%$

       2.   Original Price: = £3.15 - £0.35 = £2.80

            Make a fraction: $\frac{0.35}{2.80}$ = 0.125 = 12.5%

            For both questions the percentage increase was 12.5%

p.21   1.   Change = £9950 - £7100 = £2850 $\frac{2850}{9950}$ = 0.286 = 28.6%

       2.   Change = 840 - 630 = 210 $\frac{210}{840}$ = 0.25 = 25%

p.23   1.   0.08 x £30 = £2.40   2.   0.23 x 900 = 207
       3.   0.84 x 80 = 67.2   4.   £25 x 0.175 = £4.375
       5.   £690 x 0.175 = £120.75   But money can only be expressed to 2 decimal
                                                   places so we round up to £4.38.

p.25   1.   £250 x 1.12 = £280   2.   £25 x 0.92 = £23
       3.   £35 x 1.15 = £40.25   4.   £25 x 1.175 = £29.38
       5.   £86 x 1.125 = £96.75   6.   £145000 x 0.94 = £136300
       7.   £12 x 0.75 = £9   8.   £12500 x 0.91 = £11375

p.26   1.   Original price = $\frac{2700}{1.50}$ = £1800

       2.   Original price = $\frac{7200}{0.92}$ = £7826

       3.   Full price = $\frac{10.75}{0.86}$ = £12.50

p.27   1.   £2500 x 1.04^3 = £2812.16
       2.   £25000 x 1.05^{10} = £40722.37

p.29   1. a) yellow : red ⟶ $\left(\begin{array}{c} 2:3 \\ 4:6 \end{array}\right)$ ×2   6 pots of red needed.

  b) 1.5 pots of red.

  c) $\left(\begin{array}{c} 2:3 \\ 3:4.5 \end{array}\right)$ ×1.5   4.5 pots of red needed.

  d) 4 pots of yellow and 6 pots of red will give 10 pots of orange.

  2. Albert : Ethel ⟶ $\begin{array}{c} 1:2 \\ 420 \end{array}$ × 210

  Because 2 × £210 = £420, we need to multiply Albert's number by 210.

  1 × 210 = 210.   So Albert's wages are £210

  3. Maisy : Daisy ⟶ $\left(\begin{array}{c} 10:15 \\ £5 \ £7.50 \end{array}\right)$ × $\frac{1}{2}$ ×$\frac{1}{2}$

  Daisy's pocket money is £7.50.

p.30   1. 2 : 7 = 1 : 3.5   2. 4 : 5 = 1 : 1.25   3. 2 : 3 = 1 : 1.5

  4. 5 : 8 = 1 : 1.6   5. 3 : 5 = 1 : 1.$\dot{6}$ 6. 5 : 9 = 1 : 1.8

  7. Arm : height = 60 : 160
             = 1 : 2.$\dot{6}$

p.31   Each share = 30 ÷ (3 + 2 + 1)
             = 30 ÷ 6   ⟵ Because there are six 'shares' altogether.
             = 5

  so Ann has 15, Bill has 10 and Clare has 5 chocolates.

p.32   1. 3 eggs   2. 1 egg
     150 grams of flour   50 grams of flour
     150 grams of caster sugar   50 grams of caster sugar
     150 grams of butter   50 grams of butter
     1$\frac{1}{2}$ teaspoons of baking powder. $\frac{1}{2}$ teaspoon of baking powder.

REVISION FACT

✓ To find a percentage increase:

First make a fraction: $\dfrac{\text{change}}{\text{original value}}$

Then change the fraction to a percentage by making a decimal and multiplying by 100.

QUESTIONS

1. In the first week of term 32 people attend football training.
In the second week there are 36 people present.
What is the percentage increase?

- -

2. The price of a local bus journey goes up by 35p to £3.15.
What is the percentage increase?

Two important clues:
Find the original price.
Make sure you use the same units (pence or pounds).

- -

WHAT YOU NEED TO KNOW

Just like finding a percentage increase if you have to find a percentage **decrease**, you have to create a fraction:

$$\frac{\text{change}}{\text{original price}}$$

Example 1   A house is for sale for £250,000 but the owners drop the price to £235,000.

What is the percentage drop in value?

Step 1:   The change in price is: 250000 - 235000 = 15000.

Step 2: $\dfrac{\text{change}}{\text{original value}}$ = $\dfrac{15000}{250000}$

Step 3:   Change the fraction to a decimal: 0.06

Step 4:   Multiply by 100: 0.06 x 100 = 6
... so the drop in value is 6%

Example 2   Jeans normally cost £24.99 but in a sale they are reduced to £16.99.

What percentage reduction is this?

Step 1:   Change = £24.99 - £16.99 = £8

Step 2: $\dfrac{\text{change}}{\text{original}}$ = $\frac{8}{24.99}$

Step 3: $\frac{8}{24.99} = 0.320128051$

Step 4:   Multiply by 100: 32.01 to 2 decimal places.

... so the percentage reduction is approximately 32%.

REVISION FACT

✓ To find a percentage decrease:

First make a fraction: $\dfrac{\text{change}}{\text{original value}}$

Then change the fraction to a percentage by making a decimal and multiplying by 100.

QUESTIONS

1. A new car costs £9950, but depreciates in value after the first year to £7100.

   What percentage depreciation is this?

   ----

2. 840 people attend the opening night of a concert but only 630 attend on the second night.

   What percentage lower is the second night's attendance?

WHAT YOU NEED TO KNOW

Some percentage values are easy to find:

Examples:
1. 50% of £12 = £6   (Because 50% = $\frac{1}{2}$)

2. 25% of £20 = £5   (Because 25% = $\frac{1}{4}$)

3. 75% of £8 = £6   (Because 75% = $\frac{3}{4}$)

4. 10% of £40 = £4   (Because 10% = $\frac{1}{10}$)

The simplest way to find other percentage values:

Step 1.   Change the percentage to a decimal,

eg. 4% = $\frac{4}{100}$ = 0.04

Step 2.   Multiply the amount by the decimal.

Examples:
1.   Find 7% of £600.

7% = $\frac{7}{100}$ = 0.07

0.07 x £600 = £42

2.   Find 12% of 90.

12% = $\frac{12}{100}$ = 0.12

0.12 x 90 = 10.8

3.   Find 17.5% of £22.

175% = $\frac{17.5}{100}$ = 0.175

0.175 x 22 = £3.85

17.5% is a very special value as it is the current rate of Value Added Tax.

Most things we buy have VAT added to their value.
This is a tax that is paid to the Government.

Example: A plumber charges £28 + VAT per hour.

He works for 3 hours.

How much VAT must be paid?

$$3 \times £28 = £84$$

$$17\tfrac{1}{2}\% \text{ of } £84 = £84 \times 0.175$$

$$= £14.70$$

REVISION FACTS

✓ $50\% = \tfrac{1}{2}$ $25\% = \tfrac{1}{4}$ $75\% = \tfrac{3}{4}$ $10\% = \tfrac{1}{10}$

✓ To find a percentage of an amount, change the percentage to a decimal and multiply.

Eg. $6\% \text{ of } 20 = 0.06 \times 20$

$$= 1.2$$

✓ To find VAT, multiply by 0.175.

QUESTIONS

1 Find 8% of £30. _____

2 Find 23% of 900. _____

3 Find 84% of 80. _____

4 Find the VAT on £25. _____

5 A computer costs £690 + VAT. How much is the VAT? _____

WHAT YOU NEED TO KNOW

Look again at the example of the plumber:

He worked for 3 hours at £28 + VAT per hour.

$$\text{The 3 hours @ £28} = £84 \qquad (£28 \times 3)$$

$$\text{The VAT} = £14.70 \qquad (£84 \times 0.175)$$

$$\text{So the total bill was £98.70} \qquad (£84 + £14.70)$$

But there is a way to find this with one calculation:

$$£84 \times 1.175 = £98.70$$

So, to find a total price including VAT multiply by 1.175.

Examples:

1. I buy a digital camera for £120 + VAT. What is the total price I pay?

    $$\text{Total price} = £120 \times 1.175$$
    $$= £141$$

2. A shopkeeper decides to increase the price of dresses by 25%. The dresses normally cost £18. What is the new price?

    $$\text{New price} = £18 \times 1.25$$
    $$= £22.50$$

3. In 2004 the average price of houses in Blackdown was £164000. By 2005 the value had risen by 5%. To what value had the average price risen?

    $$\text{New value} = £164000 \times 1.05$$
    $$= £172200$$

REVISION FACTS

✓ To find the total value of a 3% increase multiply by 1.03.

✓ To find the total value of a 65% increase multiply by 1.65.

✓ To find the total value including VAT, multiply by 1.175.

WHAT YOU NEED TO KNOW

If an item loses a percentage of its value, there's a quick way to find the new value.

REVISION FACTS

✓ **If an item goes down by 12%, multiply its original value by 0.88** Because 100 -12 = 88.

✓ **If it goes down by 6% multiply by 0.94.**

✓ **If it goes down by 25% multiply by 0.75.**

✓ **If it goes down by x% do 100 – x, then change to a decimal and multiply.**

QUESTIONS

Find the new values of the items below.

1 A camera costing £250 goes up by 12%.

----

2 A coat costing £25 goes down by 8%.

----

3 A pair of shoes costing £35 go up by 15%.

----

4 VAT is added to a £25 electrician's bill.

----

5 A restaurant adds a 12.5% service charge to a bill of £86.

----

6 The value of a £145000 house drops by 6%.

----

7 A hat costing £12 is reduced by 25%.

----

8 A £12500 car depreciates by 9%.

WHAT YOU NEED TO KNOW

Examples:

1. The cost of a shirt goes up by 5% to £15.33
   What was the original price?

   Original price = $\frac{15.33}{1.05}$ = **£14.60**   Note: your calculator will show 14.6 but, because it's money, you must write £14.60

2. The cost of a necklace goes down by 25% to £12.
   What was the original price?

   Original price = $\frac{12}{0.75}$ = **£16.00**

REVISION FACTS

✓ **If a price goes up by 12%,**   Original price = $\frac{\text{New Price}}{1.12}$ Because 100 + 12 = 112

✓ **If a price goes down by 12%,**   Original price = $\frac{\text{New Price}}{0.88}$ Because 100 − 12 = 88

QUESTIONS

1. The value of an oil painting goes up by 50% to £2700.
   What was the original price?

   ----

2. A car depreciates by 8% to £7200.
   What was its original price to the nearest pound?

   ----

3. A CD is reduced by 14% to £10.75.
   What was the full price?

WHAT YOU NEED TO KNOW

Example:

£500 is invested in a bank for one year at 5% interest.
What is the value of the investment at the end of the year?

$$500 \times 1.05 = £525$$

If this money is invested for another year at 5% interest, the value will rise to:

$$£525 \times 1.05 = £551.25$$

So, to find the value after two years we could have calculated:

$$500 \times 1.05 \times 1.05$$

$$\text{or } 500 \times 1.05^2$$

REVISION FACTS

✓ To find the value of money that is invested for five years at 4% interest:

Original value x 1.04^5 ← This is the number of years.

This is the interest rate.

✓ You can use the calculator:

These keys work as the index number.

QUESTIONS

1 Calculate the final value of an investment of £2500 over a period of 3 years at 4% interest.

----

2 £25000 is invested in a building society at 5% interest.
What is the value after 10 years?

WHAT YOU NEED TO KNOW

Ratio is used to find quantities of materials in relation to each other.

For example, builders use sand and cement in relative quantities, cooks use ingredients in relative quantities.

Example: Ian makes strong mortar by mixing cement and sand in the ratio 2 : 5.

If he uses 4 bucketfuls of cement, how many bucketfuls of sand does he need?

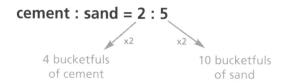

cement : sand = 2 : 5

x2   x2

4 bucketfuls   10 bucketfuls
of cement   of sand

He needs 10 bucketfuls of sand.

Look at other quantities using the same ratio:

We've multiplied   **2 : 5**   So we have to multiply
the 2 by 1.5   the 5 by 1.5

3 : 7.5

4 : 10

6 : 15 ←   Here we've multiplied
both numbers by 3.

REVISION FACT

✓ To keep the ratios equivalent, multiply both numbers by the same number.

For example: 3 : 4

and   6 : 8   are equivalent ratios.

We say that they are in proportion.

1 Yellow paint is to be mixed with red paint to make orange, in the ratio 2:3.

a) If I use 4 pots of yellow, how many pots of red do I need?

----

b) If I use 1 pot of yellow, how many pots of red do I need?

----

c) If I use 3 pots of yellow, how many pots of red do I need?

----

d) If I use 4 pots of yellow, how much orange paint will I make?

----

2 Albert and Ethel are paid weekly. The ratio of their earnings is 1:2. If Ethel is paid £420, how much is Albert paid?

Clue: the order of the names and the ratio are very important. As Albert's name was written first his earnings are related to the 1 and Ethel's are related to the 2. In other words, she earns twice as much as him.

----

3 Maisy and Daisy are given pocket money in proportion to their ages. Maisy is 10 and Daisy is 15. Maisy has £5 pocket money; how much does Daisy get?

WHAT YOU NEED TO KNOW

Look back at the last question on page 25.

The pocket money was in the ratio:

Maisy : Daisy ⟶ 10 : 15

But we can simplify the ratio just like we can simplify fractions:

10 : 15
2 : 3   Both sides can be divided by 5.

Sometimes you will be asked to simplify a ratio in the form 1 : n.

Look at the pocket money ratio:

10 : 15
1 : 1.5

Here are more examples:

2 : 5
1 : 2.5

4 : 9
1 : 2.25

REVISION FACT

✓ To express a ratio in the form 1 : n, make the smallest number become 1 by dividing by itself, then divide the other number by the same amount.

For example   8 : 15

becomes   **1 : 1.875**   because 15 ÷ 8 = 1.875

QUESTIONS

Express the following ratios in the form 1 : n.

1   2 : 7 =

2   4 : 5 =

3   2 : 3 =

4   5 : 8 =

5   3 : 5 =

6   5 : 9 =

7   Amy's arm is 60cm long and her height is 1.6m.
Write the ratio of Amy's arm length : Amy's height in the form 1 : n.
Clue: make sure you use the same units by changing the height to centimetres.

Sometimes we know a total figure and the ratio in which it is to be split!

Example:

Sam and Tariq are going to share £80 in the ratio 3 : 7.
How much do they receive each?

Sam receives 3 shares, compared to Tariq's 7 shares. This means there are 10 shares altogether, so we split £80 into 10 shares: £80 ÷ 10 = £8.

Sam's share is   3 x £8 = £24.
Tariq's share is   7 x £8 = £56.

REVISION FACT

✓ To split a total figure in a certain ratio, first add the two numbers of the ratio together, then divide the total figure by this added amount.

For example £600 in the ratio 5 : 7

Each share   = 600 ÷ (5 + 7)

                                   = 600 ÷ 12

                                   = £50

Now find each share: 5 x £50 = £250   7 x £50 = £350

QUESTION

1 Ann, Bill and Clare share 30 chocolates in the ratio 3 : 2 : 1.

How many chocolates do they have each?

- - - - - - - - - - - - - - - - - - - -

SIMPLIFYING RATIOS

WHAT YOU NEED TO KNOW

GCSE exams often contain questions about recipes.

Example: Here are the ingredients for making 12 small sponge cakes:

2 eggs
100 grams of flour
100 grams of caster sugar
100 grams of butter
1 teaspoon of baking powder

You may be asked to complete the list of ingredients to make 18 sponge cakes.

Step 1: Find the relationship between 18 and 12:

$$18 \div 12 = 1.5$$

Step 2: As there are more cakes needed, multiply each amount by 1.5

Number of eggs needed will be   2   x   1.5   =   3

This is the number that we were told we needed to make 12 cakes.

This is what we need to multiply by.

QUESTIONS

1 Use the information above to complete the list of ingredients needed to make 18 sponge cakes.

3   eggs

____   grams of flour

____   grams of caster sugar

____   butter

____   teaspoons of baking powder

2 Now find the ingredients for making just 6 sponge cakes.

____   eggs

____   grams of flour

____   grams of caster sugar

____   butter

____   teaspoon of baking powder

PASS YOUR GCSE MATHS